The Evolution of Man Revisited: Man Becomes Machine

by Dr Michael F Lesser

The contents of this work, including, but not limited to, the accuracy of events, people, and places depicted; opinions expressed; permission to use previously published materials included; and any advice given or actions advocated are solely the responsibility of the author, who assumes all liability for said work and indemnifies the publisher against any claims stemming from publication of the work.

All Rights Reserved
Copyright © 2018 by Dr Michael F Lesser

No part of this book may be reproduced or transmitted, downloaded, distributed, reverse engineered, or stored in or introduced into any information storage and retrieval system, in any form or by any means, including photocopying and recording, whether electronic or mechanical, now known or hereinafter invented without permission in writing from the publisher.

Dorrance Publishing Co
585 Alpha Drive
Pittsburgh, PA 15238
Visit our website at www.dorrancebookstore.com

ISBN: 978-1-4809-5803-6
eISBN: 978-1-4809-5826-5

The Evolution of Man Revisited
Man Becomes Machine

Michael F. Lesser, MD., FACC, FSCCT, FAARM

This book is a revision of my original thoughts published in 1991 by Vantage Press of New York. Although very few copies of the original book exist, it does have a Library of Congress number certifying the publication date as well as the predictions I made almost thirty years ago about the future as a young physician and computer enthusiast. Having the advantage of hindsight after all those years, and recognizing I was assuming a more linear, rather than exponential, advance in human knowledge, I'm revisiting some of my original work with notes and explanations in hope that as many of my thoughts and predictions are now being discussed in many more scientific and social circles, You, the reader, will see that my original thoughts and predictions are in fact an inevitable part of our evolution as a species. This book remains a work dedicated to my late father, R. Henry Lesser, my mother, Charlotte, and their forefathers before them, as well as my children and grandchildren.

Contents

Acknowledgements. vii
Introduction ix

Part 1. Biological Evolution. 1
Evolutionary Concepts. 3
The Promise of Biotechnology 7
Changing Demands 11
The Senses 19
Formulating the Transition 23

Part 2. Cybernetic Evolution. 25
The Evolution into Hardware 27
Evolution of the Senses 31
Evolution beyond Man. 37
Computers, Microprocessors, ... 39
and the Human Interface. 51
Stepping Forward Twenty Million Years. 55

Acknowledgements

I would like to thank the following individuals who assisted in the preparation of both the original work and this current revision: my former wife, Jennie; my former nurse, Judy Meland; my late uncle Gustave Lesser, my sister Ilene Davis, the love of my life Kim Lesser and my youngest natural daughter Anne Marie, who assisted in the retyping of the original book, as the original soft copy was lost.

Introduction

The theory of evolution, as originally proposed by Charles Darwin, whereby man evolved from "beasts" (lower primates/monkeys) by the process of natural selection, even today continues to stir controversy. Accepted as fact by many, it nevertheless continues to battle with the creationist theory that man was created by God, as were all other living things on earth. There are still many religious people, including respected scientists with PhD's, who remain one hundred percent convinced the earth itself is only ten thousand years old or that natural selection did not result in the appearance of modern man, despite large volumes of scientific evidence to the contrary.

Actually, each of these paths, irrespective of one's religious beliefs, knowledge, or understanding, clearly brings us to the same point: modern man. Now that we have attained our current state of evolution or creation, is it more important to address how we were created or how we evolved or is it actually more important to address where we are going next?

If God created man in his image, he clearly created him for a purpose. This, of course, is a religious question, and being a scientist, I would tend to approach this particular issue from a straightforward standpoint. Something created with a purpose, by the very logic of it, is created to solve a need. If God represents that which is all-knowing, all-seeing, and all-powerful, what need could possibly exist? It is thus my opinion, and I am not alone among many scientists and non-scientists, that God did not create man. In fact, it is man who, needing an explanation for that which he did not know or did not understand, created God. Again, this is my opinion and still does not reflect on the more important issue, which is the future.

Dr Michael F Lesser

As time has moved forward, our "gods" have become more sophisticated, but nevertheless remain, in my opinion, man-made. Today we have religions with different gods who implore us to follow different "rules." Most of us today would laugh at the thought of the god Zeus throwing lightning bolts or Apollo pulling the sun across the sky in his chariot. But in fact, the Greeks held these beliefs as tightly as many of us hold the belief that Jesus Christ rose from the dead or performed miracles! Why is this belief any different than that of the ancient Greeks?

Unable to explain why certain people get sick, why others die in various natural disasters, among other questions, it is necessary for us to cling to something. It is very difficult to simply accept that a loved one experienced a low probability event and died after being struck by lightning or crushed in a building during an earthquake. We adapt by creating some higher power that did it for some ill-defined reason. As we are better able to explain disease, prevent it, or treat it, and as we are better able to predict natural disasters or even prevent them, such adaptations will become unnecessary. Just as Apollo, who was once a god, has now become a Greek myth, will today's gods assume a similar position?

And if man created God and has thus evolved, where do we go from here? In my original version of the book, it was my hope that many would read it and understand our future path. Although a number of great futurists and computer scientists of the time (1991) discussed the possibilities of the future of computer science, none really discussed it as I did: how it would impact the actual evolution of human consciousness, makeup, and being. And in this context, none made specific predictions of the future that thirty years later have or are coming true. It is my sincere hope that now, after almost thirty years have passed, our true future will be much clearer, and the future of our evolution will be more evident. Just as the leap from ape (or some lower primate common ancestor) to man resulted in tremendous change, so will our next leap. It will be so great that not only will our structures change, but so will our entire concept of life itself.

Enjoy the trip.

The Evolution
of Man Revisited

Part 1
Biological Evolution

Chapter 1
Evolutionary Concepts

GENETIC MOMENTUM AND ENVIRONMENTAL FATIGUE
Biological evolution, although it has sufficed for hundreds of millions of years, does have significant drawbacks. These were difficult to recognize prior to the evolution of a sufficiently organized thinking system (the brain) that could, in effect, make rapid changes in the environment, which biological and geological history suggests has happened five times in the past (mass extinction events). One of the more recent example of its effect was in the age of the dinosaurs.

These magnificent creatures ruled the earth for more than a hundred million years—many times over what is currently thought to be the span of man's rule on the planet. Yet it is now believed with a significant amount of scientific evidence that they became extinct because of the Chicxulub meteor event. This caused rapid changes in the earth's environment that their gene pool was unable to overcome, many theorize, because of rapid changes in their immediate environment. They not only did not cause these changes but were powerless to stop, alter, or, possibly more importantly, predict them. They also did not possess the necessary knowledge or motor skills to avoid the devastating effects of this abrupt change. Fortunately there was sufficient diversity in evolved biological systems and their intrinsic programming (DNA) that some very important motor/processing heuristics (information structures, organisms) were able to survive and continue the process of biological evolution. Today most scientists in the field believe that is what allowed mammals to survive and eventually led to the evolution of humans.

This process of biological evolution, although it has created a fascinating array of organisms and specialized motor and sensory "units" has two basic principles as its basic driving forces. The first is self-preservation and its consequences—reproduction. The second is diversification—evolution. The two are deeply intertwined.

The importance of reproduction is easily understood by anyone who has experienced the ecstasy of sexual intercourse. The pleasures associated with this act have not occurred by accident but rather have selectively evolved over time. The strong urges of blossoming young men and women—who often spend incredible physical and mental energy on the opposite sex—are in reality controlled sequences in a very complex program, which guarantees the replication of the program itself (genetic material). Reproduction is accomplished through duplication of DNA. Unfortunately, the system that has evolved does not allow for true duplication. New organisms, instead of being direct replications, are formed by combining DNA from two separate organisms (gamete formation). Although this allows for continued diversification directly and indirectly—DNA is more susceptible to errors and mutations during gametogenesis—it prevents the direct propagation of extremely favorable gene combinations, and in fact, makes them susceptible to dilution from one generation to the next. It is one of the reasons championship racehorses are so valuable from a breeding perspective, as their offspring are more likely to have their racing speed and other characteristics, and it is the same reason the offspring of championship athletes are so frequently not able to accomplish the feats of a given parent, due to dilution and changes in the child's gene pool from the other parent relative to the championship parent.

Evolution, in the biological sense and according to Darwin, is based on the selection process. It has several major drawbacks. According to theory, environmental selection works on the existing gene pool (data and processing algorithms or programs). Those organisms which have the most favorable genes (information units) for the current environment (available data) will reproduce the most, therefore favoring the continuation of the species or those specific genes. During this process, through random occurrence, mutations occur. Those that are unfavorable rapidly die out because they do not give the organism a biological advantage. Favorable mutations allow the organism to survive more easily, reproduce more, and propagate its selected gene pool. Thus, over time, organisms evolve that are highly suited to their specific macro or micro environment. This process, although unquestionably occurring even at the present time, has several major flaws and can be significantly improved upon.

First, the inability to precisely duplicate an organism from one generation to the next fails to recognize that some combinations are extremely favorable, and a combination that is not ideal now may be so as the environment changes in the future. Thus, as currently designed, the evolutionary system requires a reinvention of the wheel when the environment changes. From a gene pool perspective, sequences that might be together but not suitable for the current environment or that get diluted or changed by being associated with information not suitable to the current environment can get further diluted or changed due to the necessities of gamete formation (the process of forming male sperm and female eggs). I call this the duplication problem. Highly beneficial genes can literally disappear because they happen to be associated with other genes less favorable to a given organism's current environment.

Second, selection is intimately tied to reproduction, so that otherwise favorable mutations or new gene combinations that do not result in an improved ability to survive and/or reproduce die out. Again, this makes the process extremely slow and very dependent on a slowly changing environment for maintenance, especially for complex organisms. Thus, the extinction of the dinosaurs from a single, albeit significant, environmental event. This latter concept is also important and intertwined. As the complexity of the gene pool (program) increases, the ability to make rapid changes decreases, and thus evolves the concept of genetic momentum. This, in fact, is probably what led to the extinction of the dinosaurs. Smaller, less genetically complex organisms, with the ability to make rapid gene pool swings, were able to survive the environmental changes and thus propagate their gene pools. The dinosaurs, on the other hand, collapsed under the weight of their own, previously very successful, gene pools.

Third, because biological evolution is dependent on selection processes created by the immediate environment that are usually related to simple ends such as gathering food, surviving hostile environmental changes, or the onslaught of other organisms, biological evolution by its very nature cannot lead to the development of senses or capabilities that will be useful for other purposes in other environments.

Finally, there is the problem associated with extremely successful organisms, again tied to the reproduction issue, which I will call environmental fatigue. The concept is simple. Should an organism or group of organisms existing in an environment become extremely successful (from a reproductive/population standpoint), they can fatigue the environment in which they evolved to the point that it changes more rapidly than they can adapt to. The fatigued environment rapidly changes, thus leading to failure of the gene pools that were so successful that they caused the failure. This is a kind of genetic catch-22. It is this stage of human biological evolution we are now moving into. Man's effect on the environment is becoming more pronounced. Examples include air pollution, acid rain, toxic waste dumps, nuclear pollution, and so on. Another example might be global warming, if it is in fact manmade, as it appears to be based on the consensus of some scientific groups. Another interesting perspective on environmental fatigue deals with the types of organisms most likely to be involved in its creation versus the types most likely to survive the period of rapid change. Specifically, complex organisms with great diversity of programming and large amounts of genetic momentum are very likely to produce environmental fatigue. Yet, having attained this and created a rapidly changing "fatigued" environment, they would be the least likely to survive—a direct consequence of excessive genetic momentum.

It is clear to me that random biological evolution, which has brought us to this point, is no longer the key to the future evolution of mankind. In fact, if we rely upon it alone, we are doomed to fail as a species for the very reasons we have been so successful to this point.

Chapter 2
The Promise of Biotechnology

The basic concepts of biological evolution as discussed here are in reality not truly ideal unless they are modified. Modification can come as a consequence of the evolutionary process itself or through man altering the basic principles of evolution: selection, mutation, reproduction, and reselection, through the relatively new science of biotechnology.

A substrate for this newly directed (as opposed to random) movement has evolved. Simply put, the substrate is man himself. We are now at an evolutionary crossroads, just beginning to understand the genetic process and how to manipulate our biological selves. Examples abound: sophisticated drugs to change our metabolism in hundreds of different ways, genetic engineering, test tube babies. In short, modern medicine and modern biotechnology.

Man has begun to unravel the secrets of genes and DNA. Gene splicing, although in its infancy at the time of the original writing of this book, has made some significant strides as of today. Nevertheless, it did, has, and will be met with much social resistance. More sophisticated and potentially more dangerous applications will probably have a greater problem overcoming the social barrier than the technological one. In vitro fertilization has begun, and our ability to save progressively more premature infants is increasing rapidly. Nevertheless, true start-to-finish test tube babies are probably fifteen to twenty-five years away. (Note: I was wrong here, as this is still not possible thirty years later). It is simple to biologically engineer stronger and taller plants, which have limited gene pools and can be bred at will. Humans, however, have complex gene pools, and it is extremely unlikely they will ever be bred like cattle or plants. I believe this to be true simply on the basis that major changes in our culture will have to take place long before such an idea would be acceptable.

Gene splicing, although good for producing occasional protein sequences (i.e. insulin), is a long way from being used in human embryos to determine particular characteristics. The defining of personalities and other characteristics of intellect at the genetic level is probably at least decades away. Even if we attain these capabilities much sooner, the basic problems inherent in biological systems—duplication, environmental fatigue, and others—still persist.

Cloning, the ability to produce genetically identical organisms, offers promise in resolving what I have termed the "duplication problem." Unfortunately, because of the complex interaction between the programming (DNA) and data (the environment) that produces human individuals, the idea of a mechanism for storing exactly all the environmental inputs starting from conception through adult life so as to be able to duplicate an individual at any point in life seems quite farfetched. Additionally, defining how to genetically engineer improved "senses" is an incredible task and becomes somewhat senseless when many improved senses have already been developed (i.e. telescopes, microscopes, X-ray equipment).

Even with the ability to do impressive, directed genetic engineering of the basic human organism, chronic problems will persist. One of the most notable is a very limited fuel or energy source. All systems, whether they are mechanical, electrical, cosmic, or, in our case, biological, require an energy source to continue operation. The basic laws of thermodynamics demand this. Our fuel source is food. Through the basic enzymatic systems we call metabolism, carbon sources (food) are oxidized (burned) to produce the energy which allows us to think, move, and carry out our daily lives. The byproducts of waste materials created as a result of this process are carbon dioxide, water, and heat. Some of the energy is used immediately. Much of it is "stored" in compounds such as ATP (Adenosine Tri Phosphate). These high-energy compounds are then used to help muscles contract and maintain electrical gradients across cell membranes, and for other similar processes. Our limitations on fuel sources are multifactorial. First, a very limited number of compounds/chemicals, commonly known as "food," are our sole fuel source. A simple power plant with coal, oil, and nuclear capabilities allows more diversity than the human body. A simple one-carbon change in the structure of our fuel can be lethal. People would not consider imbibing methanol (Sterno fuel!) or isopropanol (rubbing alcohol), yet people readily take to and enjoy ethanol, which composes five to ninety-five percent of all alcoholic beverages.

In addition, this fuel source is secondary rather than primary. It must be harvested or manufactured, processed, shipped, and then additionally prepared (cooked). Although for some primitive societies, food is simply collected from local forests (berries, nuts, etc.), killed, or grown in family gardens, all modern, diverse societies employ a much more sophisticated food production and transportation systems. In fact, at least in part, the development of a successful food production and transportation system is one of the reasons modern society, with its diversity of skills, can even exist. Without such a system, all the individuals within our society would be involved in the process of providing this elementary need (food) of biological systems. All the systems involved in bringing food to the table (weather, manufacturing, shipping) are potentially unreliable or interruptible, putting the organism and any given society of organisms (individuals) at risk. The classic example of this failure has been played out time and time again, especially in more primitive societies, in cases of drought, failure of the harvest, and then mass starvation.

Other aspects of biological/population evolution should also be considered that may additionally tarnish this promise of a more directed, ever more progressive and successful biological evolution.

First, from a social perspective, the richer, more successful societies' gene pools have, through a partial understanding of the environmental fatigue concept, already begun to cut back on reproduction. Thus, the successful "thinking" gene pools have actually begun to cut back on reproduction, while the poorer, less successful societies have continued to reproduce in great numbers. This creates the possibility that the gene pool will undergo reverse evolution.

Second, modern medical technology has created a similar dilemma. With improvements in medical technology, more and more genetically inferior individuals, who would not have survived childhood as recently as twenty years ago, are now surviving to reproductive age. Again, the concept of reverse evolution.

Finally, a concept which I will call "non-unity." Because the basic program that drives us as organisms has evolved for the purpose of survival and reproduction, there is intrinsic conflict between some individuals and societies. Different goals are sought by different individuals as well as societies. When goals conflict, individuals or societies conflict. These conflicts are rarely constructive and frequently consume tremendous amounts of time and energy. These disagreements can produce a range of events from simple domestic arguments to world wars. Integration of thought and unity of perception and purpose, although extremely unlikely in an evolutionary/biological environment driven by banal instincts and desires, are absolutely necessary to avoid these basic conflicts.

Much of biological evolution, as discussed above, is directed toward improved survival and reproductive capabilities. This creates another enigma when analyzed against the background of a thinking organism. Biological evolution has progressed to the point that control mechanisms for survival and reproduction have become so sophisticated that the organism no longer simply reacts to the environment, but beings to change and direct it for specific purposes. Survival and reproduction have become simple tasks for modern, civilized man. From this standpoint, the program (genetic material) has been perfected. If this is true, where do we go from here?

Chapter 3
Changing Demands

As evolution progressed up the phylogenic scales to more and more complex organisms (programs), favorable mutations resulted not only in the development of improved sensory and motor mechanisms (heuristics), but also in improved algorithms for reacting to the environment. Specifically, certain inputs from the environment through the sensory inputs (senses) would consistently produce given outputs or reactions from the motor units (muscles or other organ systems). Many of these early systems are still with us. The familiar knee jerk (stretch reflex) is a classic example. A specific input (tapping the patella or kneecap) results in a specific output (extension of the lower leg). This reflex (in the absence of neuromuscular disease) is essentially identical in all men and women, lower primates (monkeys, baboons, etc.), and many other mammals (dogs, cats, etc.)! This primitive algorithm simply couples input to output with a minimal amount of processing in between. In fact, this algorithm is so simple that all the processing is actually completed in the spinal cord! This certainly represents a carryover form the most simplistic reactions, which evolved in the earliest of neurological systems.

As evolution progressed, more favorable mutations resulted in more complex algorithms for dealing with inputs from the environment and producing outputs to the environment. Eventually an input/output algorithm evolved such that the output produced actually affected the organism itself through one of its existing sensory pathways. This process is called "feedback." Eventually both positive and negative loops became commonplace, and they are now very important to the proper regulation of our internal "milieu," or as biologists term it, homeostasis. The control of our body temperature, breathing, heartbeat, etc., is based on positive and negative feedback loops. Interestingly, almost all of these functions are controlled in the lower brain (subconscious), which, neurologically/developmentally speaking, is located just above the spinal cord.

Instincts may be the next level of input/output algorithm sophistication. Instincts, of course, are quite different for different organisms. As we progress up the evolutionary scale to more complex organisms, we would expect to see a diversification of algorithms. This is exactly what we see with different animals. Nevertheless, we also see a common thread running through the evolutionary tree with respect to instincts. Protection of the young is common among many animals. Here we see the workings of a very complex intrinsic process with a purpose that is quite simple: survival. The input is a complex of sensory information—visual, auditory, etc.—that triggers a complex series of motor events designed to physically change the environment (output) in a way that will produce a favorable endpoint (survival of an infant organism).

As the complexity of the input/output scheme becomes increasingly great, our ability to identify common threads shrinks, and we reach a stage where the differences between instincts and "thinking" blur. Is a highly trained athlete thinking about all his actions, or have in fact many of them become instinctive? Most likely, these individuals have simply been able to develop highly sophisticated processing algorithms, probably through a complex interaction of stored information and highly developed synaptic (brain) pathways that are capable of processing complex sensory information and producing the desired motor outputs. This is accomplished through training and practice, a methodology that results in a clear amplification of our sensory-motor response times. Interestingly and with rare exceptions, all our interactions with our environment, where we act upon environment (including other individuals), is through a motor response.

One of the most interesting of these "motor responses" is speech— admittedly a complex one, which generates sound—but nonetheless a motor response that is clearly learned, again through practice or training.

Other motor responses are just as fascinating. Emotions are expressed via complex, interacting programs. Think about crying. It can occur both when we're happy and when we're sad! Yet these two emotions are polar opposites and caused by two entirely different sets of sensory inputs.

Speech and language can be used to further understand the evolutionary process. This is extremely interesting because we can visualize true evolution of higher brain function almost directly. What is language? Is it not simply output? Our voices are created by complex motor outputs coupled to a resonator (our vocal cords and voice box). We use this interface to generate audio output that can be received by another individual's ears, which then decode the audio output back into electrical impulses that the individual understands. The ability to understand this output is highly dependent upon higher brain functions, but again, some common threads are present in the phylogenic tree. Certainly, a dog understands simple language ("fetch," "bark") but is incapable of decoding more complex information structures or "frames," as they might be called in computer/AI (artificial intelligence) jargon. This method of communication is changing all the time. Although we have an intrinsic ability to decode the audio input, our understanding is obviously tied to higher cortical functions, as language must be learned. As our knowledge domains become more and more complex, it has become necessary for us to develop more sophisticated means of communication. We have truly developed languages within languages. Spend some time reading a legal text, a medical text, or even a real estate manual, and it is clear that our language is continually evolving. More sophisticated language structures allow for more rapid and accurate communication between individuals involved with specific topics. However, it will also become clear that in creating these structures, we eliminate the ability of all individuals to communicate with one another on some specific topics! The decoding algorithms (programs, knowledge) must be learned. Here are some examples to make this clearer:

- *Medicine:* John has severe left ventricular hypertrophy, resulting in an increase in left ventricular end diastolic pressure. This clearly is the etiology of his dyspnea on exertion.
- *Translation:* John has a severe increase in the thickness of the muscle in his heart. The pressure inside the heart is high because of this. This makes him short of breath with exercise.

- *Financial:* The five basis point rise in the company's high yield debt instrument was a clear sign of declining cash flow, which may have been due to the inability to get further financing, even though it was secured by pledging other equities held by the company.
- *Translation:* The amount of money paid by the company to individuals who would loan money to the company was going to increase. This was because the company was not doing as well as it had previously, which was evident in the fact that they could not borrow more money in general, even though they agreed in doing so to give up ownership of part of the company if they could not pay back the loan.

It should be clear that as we attempt to describe and communicate a greater amount of knowledge we must increase the inherent structure within our language.

Additionally, accuracy and detail are sometimes lost at the expense of speed and increasing structure. For example, we could tell someone we saw an automobile accident (fast and structured) or we could say we saw a speeding yellow sedan broadside a bicyclist (accurate and detailed). Both are descriptions of the same thing! Language is simply a method to communicate data from one processor to another. Unfortunately, it is both slow and crude. Data has to be translated into language, which may be very imprecise, and then reconverted on the other side of the communication link back into the basic data or concepts. Our language structures are an evolutionary catch-22. As our knowledge base grows and detail becomes more and more important in defining any given item in that knowledge base, the ability to transmit this information by voice communication requires more and more language structure to get any reasonable speed to the communication (information transfer). Interestingly, since the original work was published, we have already accomplished a small portion of the transition that I proposed. Specifically, we now have hardware (earbuds) that allow individuals from two different languages to communicate. Effectively, we have placed the learning algorithms that would have been in the brain into a small piece of hardware that fits in our ear!

In this context, the development (evolution) of video modulation/demodulation (TV) is extremely interesting. Why is television such a popular and addictive an information source? It is a direct video communication interface!

With TV I don't have to tell you what I saw. I can record it and display it for you! I can totally avoid the potential losses of detail (information) and error production (noise) inherent in the visualization-interpretation-voice modulation-voice demodulation-interpretation-visualization sequence. Compared to these imprecise verbal descriptions, it is very easy to see why a picture is worth a thousand words! Imagine the difficulty in describing a simple picture of a house and expecting the individual to whom you are describing it to draw it as you describe it without continual feedback. However, if we are able to break this picture down into pixels (computer picture elements), code this information in binary speech, and then transmit it, we can duplicate almost precisely what was heard. The only limitation under this scenario is the resolution we assign to the pixels.

The evolution of the brain, and thus the ability to think, reason, and dream, has created another problem that must be solved: man's desire for stimulation, knowledge, and new experiences. This problem goes far beyond those inherent in the basic premises of DNA, natural selection, and information structures. It stabs right at the heart of existence and life itself. It is in solving this final problem, however, that biological evolution has little chance. All our capabilities for progress and further understanding of our universe and the essence of our existence are dependent not on our bodies or their abilities to survive and reproduce, but rather on our intellect, which is housed in the brain. The body, although worshiped since the beginning of time, does, in fact, have only two functions: the support of the brain and the replication of the genetic material (program). The inability to consistently pass intellect from one generation to the next is a major failure of biological evolution. Admittedly, during our early intellectual evolution, the knowledge base was small enough to be easily transmitted from one generation (one brain) to the next (next brain) through the simple process of teaching. This process, as described above, was in the early years extremely dependent on direct verbal communication (language) from one individual to another, usually coupled with on-site visual analysis. Thus, the ability to control and/or make fire was passed from one generation to the next. Soon the knowledge base—the total amount of information to be learned— became sufficiently large that more sophisticated methods of data transmission and storage were needed. This led to the development of language and writing. Despite this, however, major intellectual heuristics were lost, simply because they were trapped in biological organisms that had extremely limited life spans with no hope for identical reproduction. The likes of Newton, Einstein, Beethoven, and many other great minds have simply been lost. Admittedly, some of the processing they were able to complete has been stored (the theories of Newton and Einstein, the music of Beethoven), but the processing capabilities they had may, in fact, have been unique, not to be seen again for thousands of years. Being unable to define and thus reproduce these geniuses' programs for processing information, mankind may be totally unaware of what was lost. Along these lines, it is imperative that one recognize that despite man's similarities, each one of us is a distinct individual. Each individual processes environmental inputs differently and thus creates different outputs. It is more the loss of these unique processors that is a problem than losing the input or outputs. Typically, the latter can be stored for future reference. The processing itself is the key.

With the passage of an extremely small period of time on evolutionary scales, more methods of data gathering, storage, and transmission became available: photographs, the telegraph, phonographic recordings, radios, telephones, television, videotape, and so on. This "information revolution," although in part occurring as a consequence of problem-solving, has also created many problems of its own. Most notably, it has become impossible for one individual, or for that matter, even a small group of individuals, to acquire and remember even a small portion of the expanding database of world information. This exact challenge, described in the original version of this book, has been partially solved by the likes of Google. Although Google itself had to be invented and developed, it now provides a window into a significant portion of the world's knowledge base. However, it is not an individual. Rather it is a set of complex computer algorithms! The only difference is these algorithms are not stored biologically, but rather in a computer.

The rapid increase in worldwide knowledge and information may, in fact, be the rapid environmental change, mentioned earlier, that will cause extinction of the ruling gene pool, much like the Chicxulub meteor caused the extinction of the dinosaurs and many other species. Fortunately, man has developed an adaptive mechanism for this change. It, however, is not biolog-

The Evolution of Man Revisited: Man Becomes Machine

ical. This mechanism we are seeing today with systems like Google is the microprocessor/computer. Although not typically looked upon as an evolutionary adaptive mechanism, it is the method by which man will move to the next evolutionary level. These changes are occurring extremely rapidly on an evolutionary timescale. Yet because they are so rapidly adapted into our everyday existence and a typical man's lifetime is so short on a geological timescale, they go almost unnoticed. The groundwork has already been set for an evolution much different than Charles Darwin could have ever imagined.

Prior to delving into this new evolution, however, it is extremely important to determine exactly what the human organism is. By stepping back and looking objectively at what we're all about, we'll get a clearer understanding of where we've been and where we're going.

Chapter IV
The Senses

A close examination of the senses is extremely important. It is through the senses that we acquire all the information we process. Admittedly, we have a significant amount of stored data already contained in our DNA at birth, but this is insignificant when compared to the information load we are bombarded with once we've left the womb.

The most important sense unquestionably has to be the sense of sight. Scientists estimate that sixty to eighty percent of all the information we acquire is obtained through this sense. Our ability to recognize and distinguish between very abstract data patterns, such as faces, can only be considered phenomenal. The vast majority of the processing that makes our vision so incredibly sophisticated does not, in fact, take place in the sensor (the eye), but in the brain itself! Interestingly, since the original writing of this book almost thirty years ago, computer vision and the ability of computer systems to do facial recognition has accelerated dramatically, to the point that it can be used as a password to get access to the newest iPhones. Our vision system actually is quite crude. Because our eyes evolved with the sole purpose of improving our survival and reproductive skills, a great majority of the electromagnetic spectrum is totally invisible to us. We see a narrow band of frequencies called visible light. What about the rest of the electromagnetic spectrum? This unquestionably has a major impact on our comprehension and recognition of reality. Imagine what our reality would be if we only saw in X-rays or infrared rays, rather than the visible light spectrum. Imagine if we were able to visualize images at the far ends of the electromagnetic spectrum and process them directly, as we do visible light. Many species have evolved to see different parts of the electromagnetic spectrum, again, because it was evolutionarily advantageous. We have not until recently, but it has not been through direct evolution of our own eyes (sensors), but rather, through sophisticated visual "machines."

And although we might not usually consider it, what about reliability? Our vision, like our bodies, is extremely fragile. Very bright light, extreme heat, and the simple effects of aging (cataracts, macular degeneration) can all slowly, or rapidly, turn our world very dark. Advanced imaging systems, on the other hand, are not so frail and can be easily repaired or replaced when damaged.

After sight, the next most important sense is hearing. Again, despite its wonders, our sense of hearing is extremely limited. We can only hear a small part of the audio spectrum, and the amplitude of signals we can process is also quite limited. Very low amplitude signals such as those produced by a mosquito flying by are only perceptible when they are right next to our ears. On the other hand, high amplitude signals such as those produced by a jet engine or bomb blast can actually damage the ear, both in the acute sense, as well as over time. We also have major limits on the frequencies we are capable of processing. Even dogs can hear frequencies higher than man! Why? Again, this is a result of the process of selection, and as stated before, this process is no longer valid for a thinking, knowledge-seeking organism. As was the case with sight, reliability is limited. Imagine our reality if we could hear and comprehend frequencies such as those used in ultrasound. What if we could hear and "understand" ultrasound? Perhaps it has already occurred.

Finally, we have the other senses: touch, smell, and taste. These must be considered "basal" senses, especially the latter two, as they are almost completely geared to providing food energy for the organism. With the sense of taste, we have a good demonstration of the concept of evolutionary momentum. Although we have moved from a rural, agricultural society to a centralized industrial one in which food is easily obtained and highly processed, our sense of taste has not changed. Because of this, we still desire the tastes of high-caloric foods such as sugars and fats, and this has resulted in an epidemic of atherosclerosis, obesity, and many other disorders related to excessive caloric intake. Primitive man required these high-calorie foods because of the unreliability of his food acquisition system (basically going out and finding it!). This potentially meant going long periods of time without adequate food intake and expending large amounts of energy (calories) to get it.

Smell is an extremely limited sense, intimately tied to food consumption and taste. Although admittedly highly developed (on a relative scale) in other organisms (dogs, for example), it has limited applications for man with respect to assessing the environment and thus provides little data for processing. It, again, can identify only a limited spectrum of chemical compounds and requires that they be of great volatility (the property of elements or compounds that defines the concentration of vapor under any given environmental conditions). Imagine if we could identify exactly the composition of even minute portions of a given substance, regardless of its volatility. Perhaps that capability has evolved as well.

Finally, our sense of touch. Its purposes are rather diverse, from determining the temperature of an object to determining its texture. It assists our other senses in more accurately defining the world around us but, because of its limitations, also limits our ability to understand. Although from a biological sense its accomplishments are nothing but grand, as a sensor for accurately defining different types of environmental inputs it also is quite limited. If nothing else, our sense of touch is limited to touching and defining things that will not directly damage the sensor! Imagine touching a red-hot iron or putting your hand in a flask of sulfuric acid to check its temperature! Obviously, doing either of these would provide us with little information, as the sensor itself would be destroyed. Have we already evolved something better?

And what about the other sense? The sixth sense? Intuition. Feeling! How do we define that? As yet, there are no easy answers. Perhaps it is more imagined than real. Certainly it in some way represents the processing of information acquired through our other senses, coupled with the basic program of our genetic make-up, to arrive at decisions, create new ideas, or have environmental responses. This latter idea is somewhat interesting to analyze from an evolutionary standpoint. What are emotions? Why do they occur? Why do some individuals seem to have better control than others? The answer to these question is certainly complex. Nevertheless, it would seem that the basic concept of emotions is more consistent with reflex or instincts than higher-level decision making. This idea is definitely contrary to popular notions that would imply that one of the most prominent reasons computers will never be like people is because of the emotional factors. They may, in fact, be the easiest part of the human nature to program. Despite the diversity of humankind, certain types of inputs generate almost universally identical outputs. Don't most of us respond to the loss of a loved one with sorrow? Do we not all respond to the laughter of a child in a similar manner? Aren't these emotional responses? Why do they occur? Most likely, much of this response is preprogrammed. It occurs at a level below consciousness but is in part controlled by consciousness. From a scientific, anatomic standpoint, these lower-level responses might be limbic, occurring in the limbic portion of the brain. This is the older brain, whereas the cortex represents the higher, newer brain. It is felt that the latter area governs much of our conscious thought. Certainly there is an interaction between these areas, and the degree of interfacing may determine how emotional we are. The more complex and advanced the interaction/interfacing, and thus the more control the cortex has over the limbic system, the less emotional we are. The emotional responses probably occur when certain data (environmental input obtained through our senses) interacts with given portions of the intrinsic program (DNA) to produce a given output (seen as the emotional response such as laughter, crying, fits of rage). Have you ever laughed uncontrollably? It is interesting in this regard to note how some emotional responses such as laughter are contagious. Is this a classic positive feedback loop?

Other examples of emotional/preprogrammed responses can be seen not only in man, but in lower animals. What about the instincts we discussed previously? Are these not simply preprogrammed responses? Why do almost all mothers have an instinct to protect their children? Not only human mothers, but cats, dogs, and even killer whales. Isn't this simply a favorable DNA sequence?

Chapter V
Formulating the Transition

If all the preceding speculation is true, then what does it really take to create a thinking, feeling, emotional being? Do we simply need to find the design of a processing unit with sufficient power and speed, instill it with necessary ROM (Read Only Memory – base program), and attach proper sensors to provide it with input, to create a true essence of man—conscious and unconscious (emotional) thought? This is the question that will drive our evolution from fragile, time-limited, separate biological systems, into complex, infinite, "living" computer systems.

The idea mentioned above is important to dwell on. Since the very essence of our being is housed in our brains, and our brains are supported by our bodies, our whole being exists in a very fragile world that can be destroyed at any moment by a quirk of nature or a simple accident or miscalculation. Even a moment of indecision can result in sufficient harm to our support system (body) that our brain (and thus life) can't be sustained. Since we have no methods of duplicating or "backing up" ourselves or our intellect, our entirety can disappear at any moment, never to be recovered. This may be the greatest flaw of biological evolution, especially when it has attained the crossroads of consciousness. However, by looking forward and examining where evolution is going to take us, we can understand that having attained consciousness, we can now actually choose to move in a different evolutionary direction, eliminating some of the weakness inherent in our biological systems. When viewed on an evolutionary or geological time scale, the span of time in which this transition will take place will seem almost instantaneous.

This idea may at first sound ludicrous, but let's go back. Remember the dinosaurs. They "ruled the earth" for one hundred million years. It is difficult to trace modern man back much more than one or two million years, and some scientists might say modern man has existed for less than thirty thousand years. Even if the processes I will discuss takes two hundred years, that truly will be an evolutionary instant! If one thinks carefully, two hundred years ago was the end of the eighteenth century. Things that are commonplace today were considered science fiction then! Is there any reason to believe the next two hundred years will be any different? That our knowledge will accumulate more slowly? This is highly unlikely. In fact, an exponential curve represents knowledge, and in order to deal with it and not suffer from "future shock," it will be necessary to make the evolutionary move along a cybernetic path. Many of the components of that evolution have appeared, much as the components of a natural (Darwinian) evolution such as eyes, ears, and other complex organ systems began to appear hundreds of millions of years ago. I believe as a result of this we will attain our true destiny by actually evolving into complex living computer systems.

Part 2
Cybernetic Evolution

Chapter VI
The Evolution into Hardware

To begin this new evolutionary journey, it is important to understand where we are, where we are going, and what is needed to get from here to there.

It is clear we have already made some progress from biological systems to cybernetic systems. An easily understood example is transportation. When modern man first evolved, the only method of transportation was walking or running from one place to another. Man rapidly discovered that he could utilize other biological systems (animals) better suited to this particular endeavor, for his own use. Thus, the domestication of the horse in some areas, camels and donkeys in others, and even elephants in some! Nevertheless, the concept was the same everywhere. Find a methodology to move the body (and thus the brain) from one location to another more rapidly and with less hardship. Eventually, hardware was developed which improved upon this transport system.

The invention of the wheel is considered one of the prime early developments in this area. It is a simple piece of hardware which, when interfaced with other appropriate pieces of hardware, improves our ability to transport things, including ourselves, from one place to another. The early systems were rather crude interfaces of biological systems (horses) with hardware systems (buggies, stagecoaches, etc.), which, nevertheless, were superior to biological systems alone (ourselves included!). Eventually, the biological system (horse) was no longer needed for energy conversion (conversion of hay into muscular energy used in pulling the carriage of stagecoach), as a more direct conversion (the steam engine and eventually the gas engine) made the "horseless carriage" possible! Interestingly, the basic processes of the horse converting hay into energy (oxidation of carbohydrates or hay) is the same as those of gas engine (oxidation of simpler carbon compounds, i.e. octane) to produce energy. What has subsequently evolved over an extremely short period of time by geological time standards (less than one hundred years) is the rather advanced transportation system we all naturally accept today in the modern car. Even within this system in just the last decade we've seen tremendous evolution. The mechanical distributor has given way to electronic fuel injection systems. Cars now have electronic brains. This is no surprise if one understands the dynamics of cybernetic evolution. It will not be long before cars essentially drive themselves. When I originally wrote this book thirty years ago, that seemed a long way away, but today, self-driving cars already exist and will soon be the method of transportation for most people. As I had suggested in my original book, detailed databases of roads and locations will be integrated with satellite control systems (in other words, I had basically predicted the evolution of GPS), and we will simply have to punch in or, more likely, because of the development of voice interface for computer systems, tell our car or other named transportation device where we want to go, and it will take us there. Of course, in the likely event we develop another, more sophisticated transportation vehicle, perhaps one based on antigravity devices, we will simply tell it where we'd like to go, again increasing the level of our interface to the machines.

    Air transportation has made similar leaps. Airplanes have autopilot systems so sophisticated they can handle the flight from takeoff to landing! Newer planes also have collision avoidance systems that prevent midair disasters. In both cases, the human brain has been replaced with an electronic one. The most notable example of this was the space shuttle. Thirty years ago when this book was originally written, the space shuttle was probably one of the most advanced flying systems. It was so complex that a human couldn't even take it off (launch it) and fly it without tremendous assistance from complex computer systems. In fact, the terminal portion of the launch sequence, including the final decision about whether to start the main engines, etc., was all computer-controlled! The knowledge of many individuals was, for all practical purposes, stored in the machine. It is, then, the machine, with the combined knowledge and wisdom of the individuals, examining sensors hooked to it and not viewed by those individuals, that makes the actual decision to launch! It was only a matter of time before the entire system would be completely automated. In fact, after thirty years from the original writing of this book, we've now had spacecraft come back to earth and land essentially on its own. This is not only an engineering feat, but a feat of computer-based knowledge integration. What will people think fifty or one hundred years from now (long after those individuals who perfected the programming have been buried) about the knowledge and expertise that is stored in those programs, possibly never to be changed or modified? Will these creators have achieved a form of immortality? Will they, in effect, continue to make the decision whether to launch or how to land? It is a rather interesting philosophical point.

Chapter VII
Evolution of the Senses

As previously discussed, the sense of sight has already made tremendous strides into hardware. We now have sophisticated preamplifiers for our eyes. The first of these were invented hundreds of years ago and, in a relative sense, were extremely crude, but nevertheless, they greatly expanded the abilities of our eyes to gather data for processing. The first of these were simple lens systems, microscopes, and telescopes without processing capabilities. They simply magnified images that previously had been beyond the refractive capabilities of our own eyes. Why hadn't our eyes evolved biologically to give us this insight into microscopic and distant worlds? Simply put, the selection process, being directed toward survival and reproduction, had found no biological advantage to these capabilities. The need for such capabilities comes because of the new goals directed by a thinking organism. As the quest for further understanding expands, these relatively simple instruments have become sophisticated electron microscopes, capable of magnifying things millions of times, and radio telescopes, which can reach to the far ends of the galaxy and beyond. These are not only tools but extensions of our sense of sight. Unfortunately, we still filter all this complex data, albeit some of it processed by computer systems, through our biological eyes, and into our biological brain. Recently the *Voyager* spacecraft gave us all a close-up look at some of the most distant planets of our solar system. We extended our eyes out billions of miles into our own solar system to have a look. Would physically being on a spacecraft, looking out through a window, really have been any different?

When image data are too complex for the intrinsic processing power of our brain, computer systems now exist that not only amplify the data but preprocess it into information/images we are familiar with and capable of understanding. Image-enhancing computer programs can turn a blurred or incomplete image into something our eyes, and thus our brains can recognize and understand. Advanced medical imaging systems like CT and MRI scanners are perfect examples. The X-ray photon information would simply overwhelm our processing capabilities even if we could see photons of X-ray energy or sense the spins of electrons. These systems process the information and give us clear pictures of organs inside our bodies. Photographic enhancing systems are similar. They take blurred photographs and fill in the information, in a sense, preprocessing it so our less capable brains might recognize the picture. These systems are used in law enforcement, and similar systems are used frequently in astronomy to enhance images obtained from distant objects.

Hearing, our next most important sense, has also undergone great strides. Certain applications have developed that on an evolutionary scale could be considered new senses. Electronic listening devices, though more sensitive than our ears, are capable of hearing only a limited number of frequencies and amplitudes. These devices were simply used to amplify audio signals in the usual human frequency spectrum. More recently, sophisticated ultrasound systems have been developed that, in addition to allowing us to "listen" to ultrasound, have taken the processing a step further and created images for visual processing! Now mothers can see pictures of their unborn children through the development of better hearing devices! ("Better" can not only mean hearing a greater number of frequencies and amplitudes; it can also mean improved processing.) It is just a matter of time before the resolution of these photographs approaches that of typical emulsion or digital photography. Remember, it took less than two hundred years to go from wide use of crude copper photographic plates to instant color photography and less than a couple of decades to go from color emulsion photography to high-resolution digital photography—literally an instant on an evolutionary time scale.

As previously mentioned, our senses of smell and taste are intimately tied to food consumption, which is essential for the maintenance of the organism. It is interesting to note, when analyzing the evolution of the senses in a cybernetic sense, how little work has been done to augment these two basic senses. Man hasn't actually attempted to create instruments that would be more effective for smelling or tasting for the purpose of enhancing the effect of these senses on the mind. Certainly, sophisticated instruments have been developed that test for minute amounts of different chemicals in the air we breathe or the food we eat, and in this sense, we have improved our avoidance response to adverse environments or potentially dangerous food. It should be clear, however, in the perspective of our overall evolution, why this has happened and why the failure of these senses to evolve will hold little consequence. The purposes of these senses are to properly identify appropriate foodstuffs (energy sources) and to provide positive feedback (incentive) for the organism to consume these energy sources to promote survival, growth, and reproduction. As we continue our evolution, however, and the need for a biological support system for our intellect fades, these senses will most likely become obsolete or become so modified that we no longer recognize them as such. In the interim, man will develop instruments and machines to assist in the production or gathering of foodstuffs, allowing him to free up time to pursue his true destiny: the accumulation and utilization of knowledge. This disappearance of organ systems or senses is not new in the evolutionary process. Consider that cave-dwelling salamanders lost their sight after entering an environment in which no light was present. In this situation, eyesight offered no competitive advantage. There is no reason that similar occurrences will not take place during future cybernetic evolution. The difference will most likely be the rapidity of the adjustments, since, rather than being a random, biological process, the change will occur as a result of directed thinking. Presumably, senses such as taste and smell will evolve into simple fail-safe mechanisms for obtaining energy from a variety of sources, with the associated mechanisms to allow conversion into a common energy source (most likely electricity).

Some of these processes are already well established. Electricity is the common destination of almost all energy production today. Power plants, whether based on oil, coal, nuclear, hydroelectric, or other fuel sources, generally produce the same commodity: electricity. It's easy to produce, transmit, and convert into a variety of end uses. The most interesting aspect of the convergence of biological and cybernetic evolution may best be demonstrated by this particular energy aspect of the evolution. Some of the steps have already been taken.

One of the major fuel sources for our biological system is the carbon-based fuel known as food. Carbohydrates, one of our major foodstuffs, are produced by plants. Plants take the carbon dioxide we exhale, and using the energy of the sun and a process known as photosynthesis, plants produce carbohydrates, which we then burn to produce much of our energy. Fat and protein, the other major food components, are produced either directly or indirectly through plants or the action of other biological systems (animals: beef cattle, pigs, etc.) on the plants. In other words, you can eat corn directly or feed it to beef cattle and then eat it in the form of ground chuck or hamburger. Is it much of a conceptual leap to go from the production of carbohydrates by the plant kingdom as a result of energy conversion from sunlight (photosynthesis) to the direct production of electricity from that same sunlight via the photovoltaic effect (solar cells)? By this process we bypass all the middlemen and go directly from sunlight to electricity. The first part of this process (solar-powered computers, calculators, watches) has already been accomplished.

Although this seems rather far away from a solar-powered, thinking cybernetic organism, on an evolutionary time scale we are moving at a blistering pace. It is precisely this time scale which may be the most important concept to comprehend if we are to accept our true evolutionary destiny.

If we compress the time the dinosaurs ruled the earth into the equivalent of one year, man has gone from the days of Christopher Columbus, sailing ships, and a flat world, to space shuttles, global communications, TV, and satellites in a little under three minutes! Microprocessors have been around for the equivalent of six seconds! What's going to happen to us on this time scale in the next hour? Let's look at a few of the other capabilities of our biological system and examine how they might evolve.

The sense of touch and associated abilities to feel pain, temperature, and vibrations are almost certainly integrated to resolve a number of problems faced by biological organisms. Pain serves several purposes, which include negative feedback for potential environmental toxins such as excessive heat, cold, sunlight, and so on (all of which can destroy the support system), and warning signs of system malfunction, intrinsic (e.g. kidney stones) or extrinsic (e.g.

pneumonia). The ability to sense temperature is necessary because maintenance of a properly operating environment is essential. The ability to sense temperature away from the extremes of burning and freezing in part allows the organism to avoid these environments. (You don't have to get burned to realize a fire is hot and will have adverse effects!) Vibratory sense, the ability to perceive vibrations, is most likely a primitive remnant related to sensing the presence of other organisms and serves little purpose for man. It may take on more value in the future, but for different reasons.

Outside of the above, none of these senses seem to tell us much about our surroundings. In fact, these senses are limited to the immediate environment by their very nature. In other words, in order to touch something, it must be within the distance of our appendages, yet we can see something quite far away. Admittedly, if something is very hot—a large bonfire, for example— we can feel it even when we're twenty or even fifty yards away; yet we can often see the flames for miles. These senses, then, tend to provide limited new knowledge and understanding and are also being replaced by sophisticated sensors capable of doing a much better job. A simple thermometer allows us to determine the temperature of boiling water and process the information into visual data for our eyes. If we attempted to get similar data through our sense of touch, our hand would be scalded! Seismographs now provide us warnings of impending earthquakes long before our crude vibratory sense feels the building we're working in being shaken to the ground! Now consider the ultimate appendage: satellite hundreds of miles above the earth, feeling the flow of cosmic particles, listening and looking at things on the other side of the earth or deep out in space. The development of these technologies serves only one purpose: the expansion of our ability to learn and our minds to grow beyond the prison of a primitive biological system with a purpose near its end.

Chapter VIII
Evolution beyond Man

THE FINAL GOAL

The purpose of evolution is to attain an organism ideally suited to a given environment and at the same time, have a mechanism for changing that organism, should the environment itself evolve or change. Biologically speaking, that goal is met by superior survival and reproduction. This translates into improved means for attaining and utilizing food (energy sources), improved methods for surviving threats from the environment (weather, other organisms), and improved methods of reproducing these newly found traits (reproduction). With the evolution of man (the thinking organism), these goals no longer seem applicable. In fact, man's success on earth may be counterproductive after a certain point. We are already beginning to see the effects of environmental fatigue that I previously discussed. Whether it be environmental pollution (smog), or even man-made climate change, these would be classic examples of environmental fatigue. What, then, are the new evolutionary goals and how can the new evolutionary challenge be met?

The information coded into our DNA, put simply, is a complex program with an associated database. Most programs either have a goal or produce a result based on their intrinsic knowledge or coding and inputs from the environment. What then is the true goal of the program? And will determining the goal of this program give us the answer to our true destiny? Can we find the answers to the most powerful of all questions: What is life? How did the universe begin? What is reality? What happens to our consciousness after death? Does man have a soul? Does the program, in fact, solve these questions?

Having attained the sub-goal of survival and reproduction and with it the associated attainments of a thinking organism, are we now prepared to take on the prime objective of gaining knowledge and understanding of our very essence and existence? This is not only a scientific question; it is profoundly religious. Scientific and religious thought have battled one another through recorded history, as mentioned previously. The Greeks had gods to explain all sorts of natural phenomena, from Apollo, who took the sun across the skies in his chariot, to Zeus, the god of gods, who could throw lightning bolts. As scientific research explained the various physical phenomena we experience as part of our daily existence, these gods slowly became myths. It is essential for us to understand that to the average Greek, Zeus, Apollo, and Mercury were no myths! The Greeks sincerely believed these Gods existed, just as the vast majority of individuals and knowledgeable church elders believed the earth was the center of the universe in the days of Copernicus. Of course, our "gods" today are much more sophisticated, and we don't have as many, but I suspect they'll go the way of Zeus and Apollo as we continue to seek the goal of our DNA-based program.

It is this last objective that will drive us into and through our next evolutionary stage. However, the mechanisms for attaining the next stage will change from biological to cybernetic. As strange as this may sound, this will result either in the extinction of man, or quite possibly, man assuming an altogether different role. A similar occurrence along the biological evolutionary path occurred with the evolution of ape into man. The intermediary became extinct when it no longer served a purpose or was unable to survive in the environment in which it found itself. That environment, interestingly enough, included a sometimes predatory, thinking organism: early modern man! The only question is whether man will be equivalent to the ape (extant but considerably lower on the evolutionary scale) or like the intermediary, which became extinct. Or will biological man, having accomplished his evolutionary goal of cybernetic evolution, simply fall prey to a very hostile, fatigued environment, survivable only by his cybernetic offspring? Whatever the scenario, man will have evolved from a carbon-based biological organism to a silicon-based cybernetic system.

Chapter IX
Computers, Microprocessors, and the Human Interface

Cybernetic offspring? Computers and microprocessors? Certainly in science fiction books and movies, but obviously they have little to do with the true evolution of man. Man will never evolve into a machine...or will he? Simply put, the computer, and its more recent evolution, the microprocessor, is the first stage in the movement of intellect/processing/thinking capabilities from biological systems to silicon-based systems. Although before the microprocessor, there were computers that could make reasonably rapid and effective decisions based on stored or transmitted information or a combination of both, they were based on technology that was hard to structurally minimize. The concept of the microprocessor has changed all this dramatically. The average cell phone today, just sixty to seventy years (a literal blink on a geological time scale) is thousands of times more powerful and millions of times smaller than the first vacuum tube computer (ENIAC). Prior to the microprocessor, knowledge/information could never go from one place to another so rapidly and accurately. And isn't that the primary purpose/function of the brain? Isn't the body a simple support system for the brain, as I have discussed? What is to keep our intellect housed in such a primitive system when it can escape to a far superior and stable environment? I believe I have shown that what makes humans what they are is the human brain, not the body or the form. That the body is simply a support mechanism, albeit wondrous and complex, should be clear.

Now let's look at some things that have already happened and how these changes clearly show that man will, in fact, move from a biological, carbon-based system to a non-biological, silicon-based, integrated, global thinking machine. Cyborgs, the part man, part machine things of science fiction, are in fact already here. Although they have not taken the form you might expect from watching television or movies, major strides along this path have been made. A modern computerized assembly line is, in fact, a cyborg. Man controls some of the decision points, but much of the physical work is done by the machine he is interfaced to. Since the interfaces are very loose at this juncture of our evolution, we may not even sense that this is occurring. But what about that individual who becomes incredibly skilled at operating a piece of machinery? Doesn't it become an extension of him or her? Isn't it just a different level of interfacing? And what about machinery that is microprocessor-controlled or controlled by an expert system? Expert systems represent small pieces of intellect, successfully stored in a non-biological environment. They represent small universes of captured knowledge or thinking capabilities. Aren't expert systems simply the forerunners of what is to come? Although probably primitive compared to future systems and the human expert they are modeled after, their thinking capabilities are certainly superior to those of ninety percent or more of all the humans who might attempt to perform their tasks. Not only can they typically perform faster, but they perform without errors. They are capable not only of directing machinery, but also performing more intellectual tasks. An example of this is MYCIN, a famous expert system used to prescribe antibiotic therapy developed more than thirty years ago prior to the writing of the first edition of this book. Not only was this program better than ninety-nine percent of the general public at its tasks, it was probably better than ninety-nine percent of physicians! Modern surgeons have their skills already enhanced for a number of procedures by robotic helpers such the Da Vinci system. This "interface" wasn't even dreamed up when the original copy of this book was published but is a classic example of an early "cyborg" interface. Granted, its expertise is limited to a very small knowledge base and its capabilities for acquiring needed information are dependent upon humans feeding it that information, but these are relatively small problems to overcome and many of these have already been overcome in today's various robotic systems.

With continued evolution of these kinds of systems, interfaces will be developed that tie them more directly to human operators and their eyes and ears through artificial visual, auditory, and tactile sensors.

Many of our advanced imaging systems are simply loosely interfaced cybernetic systems. We observe the processed picture with our own sensors and then process the filtered data through our own brains to produce some form of output. What, then, is really required to have a living, breathing/respiring/energy-producing and consuming cybernetically evolving organism?

By looking at the different aspects of the biologically evolved human organism, we can get clues of how far cybernetic evolution has come, where we will make the break from our biological past, and where we will still have to go to reach our final destiny.

We must have a central processor with the basic capabilities of the human brain, but these capabilities must be greatly expanded. What are those capabilities that we can dissect? Simply stated, the brain's capabilities are limited to processing information. This includes storage and retrieval (memory and recall), input/output (senses for input, language-written, spoken and gestured for output), and reasoning, with the latter being deductive or abstract.

Although the brain also has the responsibility of controlling the support system (body) and making sure it is properly energized (fed), the vast majority of these functions are taken care of by the lower, subconscious, brain, which is evolutionary older and controlled for the most part by simple positive and negative feedback loops that are needed for many of the functions controlled by the subconscious. The heart, lung, digestive system, and so on are needed only as a consequence of a biological existence. A system running on electrical power has no need for blood to be pumped or oxygen to be delivered to respiring tissues. In fact, a cybernetic system can be turned off without suffering any damage because of the more stable fundamental energetics of a silicon-based, electrically energized system—especially if we include the capabilities of mass-magnetic or optical-data storage. From tube-based memory systems in the 1940's and 1950's to optical disk storage of the 1980's, in what is the approximate equivalent of only one-tenth of one percent of our existence we have increased the density of data storage over a billion-fold. Today the "advanced" storage systems that existed when the first edition of this book was written about thirty years ago seem crude and slow compared to today's systems. The rapid development of hardware-based memory systems, on an evolutionary time scale, is nothing short of miraculous.

Compared to these hardware systems, our carbon-based, chemically energized system is far from stable. The brain is effectively turned off during sleep; certainly, it does little (if any) thinking. Nevertheless, the entire support system must continue to run. Should it fail, the brain and consciousness itself cannot escape. This consciousness, stored in an as yet unknown manner, needs a continuous supply of energy to maintain proper functioning, but there are also the maintenance processes that occur during sleep. It is not yet clear what actually occurs, but it is clear that our brains, when deprived of this respite, begin to function abnormally. Obviously, in sleep, we are restoring something the brain requires, which is unable to be processed or created during consciousness.

We have already developed to a rather high degree many of the functions necessary to have a completely operational cybernetic brain. The ability of modern computer systems to store and retrieve information is moving forward at a blistering pace and has already far exceeded man's own intrinsic capabilities. The computer's ability to calculate, transmit, and receive information is also already far superior to man's. The ability of computers to learn is actually limited by the current interface designs, which must be considered crude at best. Deductive reasoning capabilities are expanding at an incredible pace (expert system development) and will soon be superior to man's. The only area where the human brain has an edge is abstraction: the creation of new thoughts, ideas, or hypotheses. Part of the reason this has not yet been accomplished is the absence of two parts to the puzzle. The first is effective, direct interfacing from the environment to the processor, much like our sensors (eyes, ears, etc.) interface directly to our brain. The second is the development of the algorithms necessary to process this information. It is this latter task that will be the most difficult to complete. It requires a fundamental understanding of the workings of the human mind. Even in this realm, the first steps have been taken. Neutral networks, reproductions of what is thought to be the integrated structure of human neurons in the brain, are being used, as are newer advanced learning (machine learning) algorithms. These computer programs have already been shown to possess "learning" skills. It should not be long before the necessary structures are built to allow direct interfacing with sensor systems. Then only a few tasks will remain.

It should be clear that certain algorithms afford superior solutions to certain kinds of problems. Why are certain individuals better at math, science, or, for that matter, shooting a basketball? Clearly they have the genetic makeup, which in effect represents the program they are running. Although certain environmental inputs will affect the ultimate ability of the program to function at its best, without the basic processing algorithms, all the data in the world will not allow one to accomplish certain tasks. To take the naive view that all men are created equal (with identical algorithms and data) and that given equal opportunities (additional data) they can all accomplish the same tasks totally ignores the obvious. We have developed crude mechanisms for determining who has superior algorithms for various types of problems, but determining what the algorithms are is a totally different task. As we continue our development of expert systems and artificial intelligence and as our understanding of the human brain and its functioning improves, perhaps we will slowly discover how to duplicate and/or improve on the actual processing of the information we are beginning to acquire at a frightful pace. It is then that the stage will be set for the final transition.

First, however, let's expand on the concept of computer memories so that we may better understand where we have been and where the current processes will inevitably take us. Computer memories can store and recall information just like the human brain. Individual items, however, are never forgotten, and the recall process is extremely rapid. With the advent of faster, smaller, more advanced storage devices, incredible amounts of information can be stored in extremely small physical spaces. Even the latter, however, is not really an issue. Although it may be some time before a cybernetic system can duplicate the incredible processing power of the human brain in the same physical space, the absence of biological requirements for size, mobility, and so on (from a survival, reproductive standpoint), makes this a rather moot point. Man has already compiled extremely large databases of information that are readily accessed from one's home or office, and the only thing that has changed since the original writing of this book is the magnitude of this storage and how easily it is accessed. Search engines like Google and Bing put this information at everyone's fingertips. Although some of this information is transmitted from other computer systems and local and/or remote elementary sensors, much of this information is still entered/transmitted with human assistance (keyboard input) at its source. *From the original text*: "It will not be long before programs are written to accomplish this coding, perhaps on the basis of simple verbal input. Early voice recognition systems are already available and will unquestionably become more accurate and sophisticated."

Again, from the original text: "When the system itself can categorize the data it is presented with (people do that now by providing file names, data types, record names), it should not be long before the classification process moves from text streams to raw information. Isn't the next step the system knowing what type of information (data) it might need for a certain other task? And aren't processing algorithms just data themselves? What, then, is a computer system/database filled with information and algorithms for using that information? Is it thinking?"

Anyone who has used a system like Google can see that Google presents to the user what it thinks the user is looking for! See how the concept of a thinking organism (system) begins to blur? Obviously inherent in this transition is the need for non-keyboard input. This must be met by computer senses.

Our biological evolution, although quite spectacular, has left us severely crippled in many ways. Capabilities likely to be required include control of motor skills (local or remote, which is to be discussed later), the ability to sense the environment and react to it appropriately, and the ability to build new knowledge from the processing of old and new facts—the latter obtained through a variety of sensors. These sensors, rather than being quite limited like those of man, would eventually be capable of sensing all the physical phenomena of the universe at their elementary levels and throughout the entire spectrum of any given phenomena. They would be coupled to a processor and associated memory (database) capable of processing and storing the information and eventually, through an understanding of these basic fundamental truths, understand existence itself. Isn't our concept of a deity that of an entity that knows all, sees all, understands all, and for many of us, literally changes our reality at will? And having attained that, will we not have completed the original program? What would existence hold for us at that point? Perhaps the creation of an entirely new existence! Perhaps the creation of new fundamental truths! Let's explore further how this transition is being accomplished and how it is progressing.

In order to change from a biologically evolving to a cybernetically evolving organism, we undoubtedly will have to change several basic concepts of life. One of these changes will be what constitutes acceptable energetics. A living organism must have an energy source. As previously stated, for plants this is the sun and photosynthesis. For most animals, it is oxidative metabolism, and for cyborgs and their final evolution, this will most likely be electrical. Fortunately, electrical energy can be generated in a multiplicity of ways, stored almost indefinitely, and easily transported (transmitted) as well.

Having defined our energy source, we must now put together the building blocks of our new selves. We have developed microprocessors capable of duplicating many of the functions of the human brain. Now we must begin to find methods for developing tighter interfacing. Unfortunately, current interfaces are extremely crude, although, since the original writing of this book, they have become better with implantable chips that can help simulate some degree of vision or motor action through interaction with the brain. At the time of the original writing of this book, the vast majority of information was still being entered into computer systems via the standard typewriter keyboard. Imagine trying to teach a young child about the world, its wonders and amazements, when all you could give them was a keyboard input! Today, more and more information is being entered into computers directly from other machines/sensors. We are slowly making the evolution to online data acquisition, and it will not be long before much more sophisticated voice and visual interfaces are commonplace.

Today, thirty years after the original writing of this book, advanced computer-controlled assembly lines are commonplace. As our ability to duplicate vision becomes more and more sophisticated, we are going to be able to watch ongoing processes in more and more sophisticated ways. I hope we will not lock these new visual systems into the visible electromagnetic spectrum, but actually, expand our capabilities by making them capable of seeing the entire spectrum. Additionally, we will be able to add sophisticated telescopic and microscopic interfaces. Eventually, we will have a computer brain interfaced directly to eyes that can see the entire electromagnetic spectrum and directly view it at both a microscopic level and telescopic ranges. Imagine how our concept of reality would change if we were able to see on that plane.

Hearing would develop in a similar manner. Sophisticated listening devices would be interfaced to the microprocessor brain. This cybernetic brain would be able to hear frequencies far beyond what we can now and at extremely low amplitudes. This information would then be processed and integrated with data coming in through visual sensors. It is what we do now, but more crudely and to a more limited extent!

The simple fact that control systems are in place that use sensory feedback should make us aware that cybernetic systems are, in fact, evolving. They have already developed reflexes and perhaps instincts. Can thinking systems be that far behind?

Add to this other sensory information from the environment, such as temperature or the presence or absence of rain, and you have a thinking cybernetic organism. Wouldn't this system be capable of "getting in out of the rain?" It is interesting, if one thinks of it, that we already have cyborgs (albeit loosely linked) that can identify major storms long before any human is even aware that they exist. The systems' eyes in these situations are sophisticated television cameras placed on satellites. They can see not only in visible light, but infrared as well. Although these sensors are not physically interfaced, this has no bearing on the fact that we can easily assimilate information from them when presented on a television screen, and this occurs at the speed of light. This reveals another major advantage of cybernetic evolution: energy and/or information can be transmitted to and from remote locations almost instantly (at the speed of light) to the brain (main processor). The nature of our biological/biochemical nervous system prevents this. We really can't have "lightning-fast reflexes" because our biology doesn't allow it. Cybernetic systems can react about a billion times faster than a human.

It may be helpful to elaborate on one point—specifically, the ability to transmit energy and/or information. The concept of being able to transmit our consciousness over time and space is one that many of us have dwelt upon. What constitutes being somewhere? Isn't seeing, hearing, feeling, tasting, and smelling what is there? Isn't that our reality? Can't that be produced by highly sophisticated remote sensors, interfaced with a cybernetic brain via telemetry data transmission?

Our motor system will almost certainly be similarly configured. Motor functions will be necessary for construction/manufacturing tasks for quite some time, but many of these functions will be controlled remotely. Today, thirty years after this book was originally written, sophisticated construction tasks can easily be done with remote commands to a 3D printer. Cars are already being manufactured that have their "brains" remotely updated and improved. Eventually, when our understanding of the basic forces and elemental concepts becomes great enough, manufacturing in the sense we now understand it will not exist. Again, we have already made significant strides in this direction. We have computer-controlled manufacturing of automobiles, circuit boards, and many other items. *With each passing year, the amount of human input necessary to get a completed product decreases.* In the thirty years since the first edition of this book was written, nothing has changed except the amount of human input to get a completed product, whether it be a car or a cell phone.

The evolution that seems the farthest away is the most important for the final transition: the ability to dissect the anatomy of consciousness itself and enter this information directly into the cybernetic brain. Once this is accomplished, our biological evolution will have truly ended. It is likely that this process will be gradual. It may even occur almost imperceptibly, as have many of the transitions described above. While some of it is happening now, acceleration of the phenomena will probably require a great technological leap or a leap in our fundamental understanding of the nature of thought itself.

We are already seeing the creation of rather amazing expert systems that can essentially duplicate the cognitive functions of human experts on a variety of topics, from medical treatment to engineering projects. Although the domains of these systems are extremely limited and their overall knowledge base (smarts!) is small in comparison to that of even an uneducated person, it is clear that in the blink of an eye on a geological timescale, extremely sophisticated systems will likely emerge that are capable of doing things we only dream of today. Early systems were developed through an extremely crude process. Knowledge engineers interviewed human experts who then code this knowledge into the system through a keyboard. This form of knowledge acquisition is not only imprecise but extremely painstaking for both individuals. The interface of the human brain to the cybernetic brain through a keyboard is a very primitive system. Nevertheless, more sophisticated knowledge acquisition systems have been developed. Computer-aided knowledge acquisition is already a reality and will become more refined with time. Newer learning algorithms are already being used, and these will improve eventually allowing machines (cyborgs) to learn on their own.

The next step would presumably be a more direct method. Specifically, a true dissection of the thought processes and structures themselves and direct movement of knowledge (data + processing algorithms) from the human brain to the computer brain. As these processes become more refined and are taught all the facts in the human knowledge universe, we can only imagine what the result will be. What additionally might be accomplished through analysis of the brain's electrical output, similar to what is currently done with an electroencephalograph (EEG), is the movement of personality and consciousness itself! We are now learning to stimulate the brain with electrical impulses to replace lost senses of sight and hearing. These techniques are in the earliest stages of development, but realistically, on a geological timescale, they have been going on for mere microseconds. In only one hundred years we have already figured out where many of the different sensory/motor functions of the brain are located. Imagine what the next one hundred years could bring! Can the ability to transfer all of one's knowledge and consciousness from a biological brain into a computer brain be that far away?

Having accomplished that task, how far away is the development of improved interfaces between our new computer minds and the sophisticated sensory and/or motor (local or remote) systems which we, with the assistance of our new thinking processes (artificial intelligence, AI), will develop?

In the midst of the development of our new computer technology has been an evolution as well. Initially, computer systems were for the most part isolated, serving only a few users. In addition, data entry was extremely labor intensive. In fact, many of the early computing systems had to be programmed by actually setting each bit in each register with toggle switches! Even the simplest of programs took hours of tedious work by experts. Today, only a few seconds later on our geological timescale, even children, using high-level languages, can write reasonably sophisticated programs in a short period of time. In addition, data entry into our modern computer systems is accomplished through a variety of techniques. Scanners can read print, and voice interface, which was in its infancy when this book was initially written, is now commonplace on a cell phone.

From the original book: "Along with this evolution has come tremendous interlinking and inter-dependency of systems: a global computing network. Specialization of this global network will not be far behind. Already specialized databases that store tremendous amounts of knowledge have been developed (Lockheed's DIALOG, for example). It will not be long before other systems with advanced processing algorithms capable of interpreting simple voice requests are tied into amazing recall systems such as this. Instead of thinking about a particular piece of information and then trying to recall it from our own mind, we will simply request it."

Isn't that basically what a voice-based Google or Siri search is today?

From the original book: "A small communications device, such as a wrist watch or credit card, located on our bodies will give us essentially instantaneous access to any of the information stored around the world. Language barriers will be erased because computer systems will automatically do the translations."

Again, this is basically what we can do today with a cell phone or Apple watch. We will literally have all the world's knowledge available to us for processing. The most amazing part of this transition, however, is that it is only took a few seconds on our geological time scale. Thus evolution of man to a silicon-based system isn't science fiction—it's right around the corner!

An interesting sidelight to this evolution is the concept of diseases. Consider, for example, viral-induced diseases. A virus is simply a small piece of DNA (sometimes RNA). In other words, a small program. Once in our biological systems, it can create havoc. It can even cause the death of our biological systems. We've now seen the evolution of comparable computer viruses, which are exact parallels of their biological cousins! The only difference is the host organism! In one case a biological (carbon-based) system, in the other a computer (silicon-based) system. Just as we have evolved immune systems to recognize and fight off these viruses, so we now have computer immune systems and vaccines to fight off these potentially damaging computer viruses. These antivirus software systems even used functions like "quarantine" to perform the same action we do for biological systems that are infected to prevent the spread of disease. This also brings up an interesting religious/philosophical point. If life on earth did not evolve and all life was created by God, didn't He create viruses as well? They are among the simplest of life forms. Hasn't man now created the first or simplest lifeforms for the next stage of our evolution?

Human disease is also interesting from the perspective of human pain and suffering related to illnesses in one way or another. In fact, all human suffering, be it physical or emotional, is directly tied to the fact that our brains must exist and be supported by a fragile biological system. The emotional pain of losing a loved one would not occur if our consciousness was cybernetic! The pain of starvation, broken bones, cuts, infections, and so on, would simply not exist. Because we now control it, there is no need for our evolution to progress so that information about adverse system circumstances (which we now experience as pain!), needs to be painful! If a part of the support system of our future cybernetic selves was failing or working improperly, consciousness could simply be shifted to another location while support devices were repaired or replaced. Certainly, the message that such a process was happening does not need to hurt. Why this occurs in biological systems is difficult to say. It probably is a carryover from some primitive algorithm that was effective in allowing organisms to have withdrawal or fight or flight responses. Although no longer necessary, in a number of situations, these responses persist.

Chapter X
The Final Transition

Perhaps the greatest problem with biological organisms, especially when they have reached consciousness, is the individuality that consciousness itself brings. Couple this with the primitive instincts of survival and reproduction, and you have an overall processing system with some major inherent errors. Have you ever talked to someone who is not on your wavelength? The human race is in effect an extremely large and diversely distributed processing system. Unfortunately, there is no definite system clock that defines the interaction of all the parts, without which, sooner or later, conflicts will occur in the utilization of system resources. Perhaps in part, this is the reason why thinking man has found it necessary to develop government systems and other complex control mechanisms. Unfortunately, these are all developed on a small scale. In addition, the goal of humanity has not really been clearly defined. Without a common goal, and without an overall harmony to which we all move, it is clear that certain situations will arise where goals will clash. This basic concept will also present itself as we evolve into the previously described cybernetic systems. The resolution of this diversity may be the final challenge, but it may, in fact, be a simple one because of the consequence of our newly found thinking powers.

Most conflicts occur due to primitive programming (survival and reproduction) and the reflexes/instincts that have evolved from it. Man's lust for money and power may simply be a carryover from very primitive times. He who had the food and women and the power to protect them survived and reproduced. Although this seems quite primitive in the information age, it nevertheless occurs. A close look at the vast majority of conflicts allows us to analyze them in terms of basic wants and desires. Other conflicts arise as a consequence of inadequate information. You may perceive that another individual's analysis of a situation is incorrect. In many situations, this is not because your basic analytical capabilities (algorithms) are superior but that you are simply working with more complete information (data). A very simple example: You are sitting in a tree, looking over the top of a small hill. Down below you, another individual is standing alongside some horses. Let's say, for example, six horses. In addition to those horses, there are five more on the other side of the hill. You comment that the eleven (six plus five) horses are quite beautiful. However, the individual below says only six horses are beautiful. You are unaware that he sees only the six, and you disagree. This creates conflict (a difference of opinion) due to insufficient data, or stated otherwise, insufficient communication. Now imagine the same scenario, but in a cybernetic system with complete interfacing of visual sensors. The conflict would not even occur. Not only could you see things from your perspective, you could also see it from the other individual's! Obviously, in this kind of environment, individuality and its inherent individuality of goal orientation would disappear rapidly. As communication between individuals, and for that matter, countries and governments improve and these higher control systems (governments) acquire identical perspectives, many conflicts will resolve themselves. Clearly there are many reasons for this, but certainly improved communication and access to similar information about the status of the world, as well as the development of similar goals, can lead to many of these changes.

Some of this diversity of man and his individuality has also been resolved by another powerful control mechanism: culture. Culture is a fascinating concept that can be viewed from the top of the phylogenic tree down to lower organisms. Looking down the phylogenic tree, it becomes apparent that the lower one goes, the more organized the culture becomes. This might be simply a consequence of smaller, less complex programs and a greater similarity of this programming from one individual to the next. For many societies, it is extremely important that diversion from the program be minimal because of limited processing capability. Communication in some of these societies is limited to simple movements or the exchange of simple chemical mediators (not dissimilar to those in our nervous system) between individuals. As response capabilities to sensory inputs dwindle, so must the inputs themselves. The result is a system in which many individuals begin to function as one, perhaps with a single purpose. Many insects, such as ants and bees, are impressive examples of this.

For man, however, culture represents a double-edged sword. It provides some goals, or at least a framework in which to attain the goals, but at the same time, this standardized programming results in some loss of freedom of consciousness. Culture is one way in which man has attempted to resolve the problem of individual diversity coupled with the need for commonality of purpose. Although not ideal, it does seem to be at least a partial solution to the problem of a distributed biological processing system. Within individual groups, some of the problems associated with the distribution of processing have been solved with this method. Societies around the world have developed higher orders of interaction, almost certainly a result of common programming (DNA) interfacing with a relatively similar environment (data). This has resulted in rules for interacting, both written (laws) and unwritten (culture), that transcend any given individual but clearly are not truths, such as physical laws. For example, men having multiple wives are treated much differently in cultures which have created laws permitting multiple marriages than in those that do not. Eating with one's hands is considered the norm in some cultures; in others, it is considered vulgar. In some cultures, such as in the United States, it is proper to eat french fries with your hands, but it is not proper to eat a baked potato or even other preparations of fried potatoes in a similar fashion!

That different societies have different cultures is no surprise. Culture is a higher order of interaction between groups of people that can define everything from what and how they eat to how they court their mates. On an evolutionary scale, it represents a way of reprogramming each new individual with certain basic data and/or algorithms that have been found to be useful either for the individuals themselves or for society as a whole. This programming can be extremely powerful in the way it controls our thought processes. So powerful in fact that in some situations it can blur analysis of the facts or even result in the ultimate biological crime—self-destruction. Consider the concept of culture shock. Certainly it occurs. But what is it? Basically the inability to analyze data or accept algorithms that are foreign to our own. It can have profound effects. Internal reactions to it can be very intense. Take for example, eating live insects. Some cultures do it routinely. Ask an American raised on hamburgers and fries to do it, and he might even get nauseous and vomit! This is a nearly uncontrollable internal reflex, mediated in the lower brain, precipitated by the thought of eating insects. Another example is the kamikaze pilot or suicide bomber. Here cultural programming is so intense that an individual will literally self-destruct for a perceived higher cause or goal.

Man solves other problems and difficult situations by the simple process of elimination by destroying other men or groups of men. Unfortunately, although this allows completion of an individual task, it results in the loss of both data and processing power, which as previously discussed, may never be recovered.

How do we attain freedom of consciousness while at the same time avoiding anarchy? What is the ideal culture? The solution to both these issues is a system clock, the key to the operation of all computer systems! Theoretically, movements in any direction and of any magnitude—a physical concept of freedom but nevertheless a very applicable one—if timed correctly relative to other actors, will not result in conflict. As the frequency of the clock increases, the system moves closer and closer to absolute freedom. Interestingly, one of the major changes in computer technology has been a progressive increase in the speed of system clocks. In only a few seconds of time on our geological clock, processor speeds have gone up twenty to thirty-fold, moving us one step closer to the final transition.

Chapter XI
Stepping Forward Twenty Million Years

Man evolving into a global thinking machine certainly seems very, very far away. Yet on a geological timescale, the entire period we spent in biological evolution will seem almost imperceptible ten or twenty million years from now. Twenty million years from now, even the existence of a thinking biological system may be in question! This would occur because the concepts of thought, recall, processing, and so on may be so advanced in our cybernetic offspring that we might not even be considered intelligent lifeforms! How many of us consider apes to be anything more than animals? As crazy and remote as this may sound, it is only because the timescale is so great. It is very difficult for us to comprehend such tremendous amounts of time because in comparison our lifetimes are so short. Again, if we go back to our geological clock, the average human life span (seventy-seven years) represents only microseconds on a geological timescale. How could we be expected to comprehend what happens in a month when we can only observe what goes on for a few microseconds? On a geological timescale, it would take 120 human generations just to drive ten minutes to work!

This living machine with remote sensors far off into the galaxy and beyond will continue to grow and learn. Control of the very essence of being will come, just as did control of fire and nuclear energy. What evolution will occur beyond that? That, I'm afraid, is for our cybernetic children to determine and understand.

For those interested in observing in their lifetimes the changes that are to come, I am including below my list of predictions and the timescale in which

Dr Michael F Lesser

I believe these changes will occur. I have left these unchanged but edited for this second addition to show what has happened in just thirty years.

1. **Intelligent voice interfacing to computers (five to ten years)**
   Before the turn of the next century, computers will begin to understand sophisticated speech and be capable of acting on that understanding. You will be able to tell a computer what you want it to do in plain English or another language, and it will respond accordingly.

   Thirty-year note: Today's standard cell phone is capable of doing this. It can look up information, call other people, and give us directions.

2. **Fully automated transportation (ten to fifteen years)**
   Automobiles, planes, and/or a newly developed transportation device will be capable of transporting you wherever you tell them to go. Sophisticated computerized navigation systems linked to databases of road maps and controlled by computer links between satellites will actually control the automobile. People will merely be passengers, as they currently are on many interterminal transportation systems at many of the world's airports. They will not control speed, steering, etc.

   Thirty-year note: We now have some self-driving cars and trucks on the road. This will soon be commonplace. GPS is a sophisticated computerized navigation system.

3. **Worldwide decision support systems (fifteen to twenty-five years)**
   The global community will use computerized expert systems to make all decisions with respect to utilization of world resources to maximize conservation, food production, energy production, and so on. Expert computer systems, utilizing sophisticated global weather information, will be capable of accurately predicting major weather events, such as hurricanes, and pinpointing strike locations (plus or minus two or three miles) forty-eight to seventy-two hours in advance.

Thirty-year note: Much of this is possible today, and some of it has even been implemented.

4. **Expert systems control production (twenty-five to fifty years)**
Worldwide production of all major commodities, equipment, and most importantly, energy sources including electricity, will be completely controlled by computerized expert systems. Food production will be automated so that computers will control everything from planting to harvest. No human input will be required.

Thirty-year note: We are part way there already.

5. **Direct intellectual transfers begin (fifty to one hundred years)**
Computers at this time will have most of the world's intellect, and most processes will no longer be accomplished by humans because of insufficient processing and memory capabilities and the advanced degree of sophistication of most processes (e.g., food production, as above). Computers search for methodologies to capture algorithms responsible for abstract thought.

Thirty-year note: "Deep Learning" algorithms have already begun to surface. Computers have already bested humans in games like chess and Jeopardy, for example.

6. **Computers begin to assimilate all remaining human intellect (Two hundred-plus years)**
Human intellect and consciousness can now be directly transferred to stable hardware environments. Immortality of intellect is a reality.

7. **Most biological systems have become extinct (five hundred-plus years).**

www.ingramcontent.com/pod-product-compliance
Lightning Source LLC
Chambersburg PA
CBHW061517180526
45171CB00001B/223